MATERIALS SCIENCE AND TECHNOLOGIES

^{57}FE MÖSSBAUER SPECTROSCOPIC AND DENSITY FUNCTIONAL THEORY (DFT) STUDY ON THE INTERACTIONS OF THE METAL ION WITH MONOSACCHARIDES

MATERIALS SCIENCE AND TECHNOLOGIES

Additional books in this series can be found on Nova's website under the Series tab.

Additional E-books in this series can be found on Nova's website under the E-book tab.

MATERIALS SCIENCE AND TECHNOLOGIES

^{57}FE MÖSSBAUER SPECTROSCOPIC AND DENSITY FUNCTIONAL THEORY (DFT) STUDY ON THE INTERACTIONS OF THE METAL ION WITH MONOSACCHARIDES

YASSIN JEILANI
BEATRIZ H. CARDELINO
AND
NATARAJAN RAVI

Nova Science Publishers, Inc.
New York

Copyright © 2011 by Nova Science Publishers, Inc.

All rights reserved. No part of this book may be reproduced, stored in a retrieval system or transmitted in any form or by any means: electronic, electrostatic, magnetic, tape, mechanical photocopying, recording or otherwise without the written permission of the Publisher.

For permission to use material from this book please contact us:
Telephone 631-231-7269; Fax 631-231-8175
Web Site: http://www.novapublishers.com

NOTICE TO THE READER

The Publisher has taken reasonable care in the preparation of this book, but makes no expressed or implied warranty of any kind and assumes no responsibility for any errors or omissions. No liability is assumed for incidental or consequential damages in connection with or arising out of information contained in this book. The Publisher shall not be liable for any special, consequential, or exemplary damages resulting, in whole or in part, from the readers' use of, or reliance upon, this material. Any parts of this book based on government reports are so indicated and copyright is claimed for those parts to the extent applicable to compilations of such works.

Independent verification should be sought for any data, advice or recommendations contained in this book. In addition, no responsibility is assumed by the publisher for any injury and/or damage to persons or property arising from any methods, products, instructions, ideas or otherwise contained in this publication.

This publication is designed to provide accurate and authoritative information with regard to the subject matter covered herein. It is sold with the clear understanding that the Publisher is not engaged in rendering legal or any other professional services. If legal or any other expert assistance is required, the services of a competent person should be sought. FROM A DECLARATION OF PARTICIPANTS JOINTLY ADOPTED BY A COMMITTEE OF THE AMERICAN BAR ASSOCIATION AND A COMMITTEE OF PUBLISHERS.

Additional color graphics may be available in the e-book version of this book.

LIBRARY OF CONGRESS CATALOGING-IN-PUBLICATION DATA

Jeilani, Yassin.
 Interactions of the metal ion with monosaccharides / Yassin Jeilani, Beatriz H. Cardelino and Natarajan Ravi.
   p. cm.
 Includes index.
 ISBN 978-1-61122-301-9 (softcover)
 1. Organoiron compounds. 2. Monosaccharides. I. Cardelino, Beatriz Hoffmann. II. Ravi, Natarajan. III. Title.
 QD412.F4J45 2010
 547'.05621--dc22
                    2010036146

Published by Nova Science Publishers, Inc. † New York

CONTENTS

Preface		vii
Authors' Contact Information		ix
Chapter 1	Introduction	1
Chapter 2	Experimental Methods	5
Chapter 3	Computational Methods	7
Chapter 4	Mössbauer Spectroscopy	9
Chapter 5	Results and Discussion	13
Chapter 6	Properties Related to Mössbauer Spectroscopy	29
Chapter 7	Conclusions	37
Acknowledgment		39
References		41
Index		45

PREFACE

The fundamental electrical and magnetic interactions between iron ions and biomaterials were investigated using an experimental approach, ^{57}Fe Mössbauer spectroscopy, and ab initio computational methods. A conventional spin-Hamiltonian approach adopted for the data analysis of the Mössbauer data showed that the metal ion in the Fe-chitosan complex is in the high-spin ferric state and that it has an internal magnetic field of approximately 440 kG, at the nucleus. The magnitude of the internal field arises from the predominant Fermi-contact interaction of the high-spin ferric species with N/O ligands. This book proposes a scheme for the Fe-chitosan complex based on the analysis of the experimental data.

Authors' Contact Information

Yassin Jeilani
Environmental Science Program
Spelman College
Atlanta, Georgia 30314, USA

Beatriz H. Cardelino
Chemistry Department
Spelman College
Atlanta, Georgia 30314, USA

Natarajan Ravi[*]
Physics Department
Spelman College
Atlanta, Georgia 30314, USA

---

[*] Corresponding author; email: nravi@spelman.edu

Chapter 1

INTRODUCTION

Biomolecules such as cellulose, glucose, chitosan, chitin, and chondritin sulfate form very stable complexes with a variety of metal ions, especially transition metals. These metal complexes have a wide range of applications, for example, in sewage system purification or in cancer research. Cellulose is a polymer of glucose with β(1→4) linkages ideal for making fibers since parallel polysaccharide chains can be readily linked by hydrogen bonds (Figure 1). Chitin, the second most abundant polysaccharide after cellulose, is the fibrous component of the exoskeleton of several arthropods: crabs, spiders, and insects [1,2]. This polymer, which resembles cellulose structurally and chemically, offers diverse applications due to the presence of an acetylamide group at one of the carbon atom positions. Chitosan, the N-deacetylated chitin, is similar to chitin but easier to work with due to its increased solubility in acetic acid [1]. The polar nature of the carbohydrates of these biopolymers, gives them the ability to chelate with metal ions. Such metal complexes structurally resemble the active sites of metalloenzymes due to the presence of O/N/S ligands [3]. The two polymers chitin and chitosan also act as substrates for the enzymes chitinase and chitosinase, respectively [4].

Chondritin sulfate, on the other hand, is an alternating copolymer of β(1→4)-D glucuronic acid and β(1→3)–N acetyl-D-galactosamine, which can be sulfated at C^4 and C^6 positions (labels for carbon positions shown in Figure 1). The presence of carboxyl (COOH) and sulfate (SO_4^{2-}) groups makes this polymer highly soluble in water. Chondritin forms networks with collagen in connective tissues and allows the transfer of globular proteins, stimulates the growth and differentiation of neurons, and has also been implicated in spinal cord injuries [2,5]. These polymers have been found to possess an excellent ability to chelate an array of transition metal ions, even though chitosan appears to be a better chelating agent [3]. This could be due to the presence of the amine (NH_2) group in chitosan.

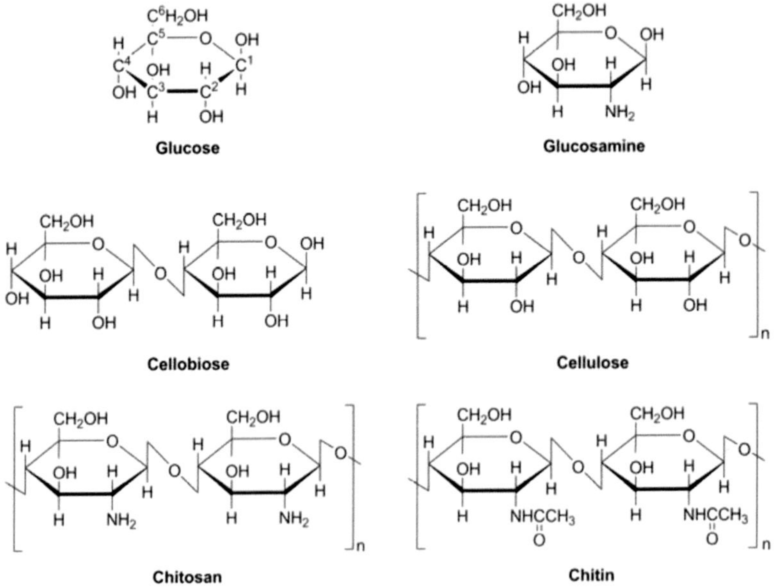

Figure 1. Chemical structures of glucose, glucosamine, cellobiose, cellulose, chitosan, and chitin.

The structural properties of transition metal complexes of chitin and chitosan have been studied to try to understand the mechanism of enzyme chelation [6,7]. Different models for the metal complexes of chitin and chitosan have been proposed. In the "pendant model" the single metal ion is bonded to an NH_2 group of chitosan, while in "the bridge or chelating model" the metal ions are proposed to be coordinated to several amino groups in the same or in different chitosan polymer chains [8]. Studies of Cu and Fe metal complexes have indicated that both NH_2 and hydroxyl (OH) groups may be bonded to the metal ions, and that more than one polymer chain may be involved in the formation of the complexes [9,10]. However, the way the metal binds to the carbohydrate matrix is still not clearly understood and hence requires further investigation.

Elucidation of the electronic and magnetic properties of a metal ion in metal complexes, metalloenzymes and synthetic analogues, have been pursued by probing the metal site with a variety of spectroscopic techniques

[11]. Some of the magnetic resonance techniques, including Nuclear Magnetic Resonance (NMR), Electron Paramagnetic Resonance (EPR), Nuclear Quadrupole Resonance (NQR), and Mössbauer effect, are widely used [12]. For example, the Mössbauer effect has played a decisive role in understanding the structure-function relationship of several Fe containing redox enzymes such as hemoglobin, cytochromes, cytochrome c oxidase, ribonucleotide reductase, hemerythrin, nitrogenase, hydrogenase, and ferredoxin [13]. The experimental studies have often been intertwined with quantum mechanics or other simulations to model the experimental behavior [13].

In order to understand the nature of bonding, geometry, and magnetic characteristics of the Fe complexes with biopolymer materials, we have undertaken a Mössbauer study, since the technique is sensitive to hyperfine interactions of the order of 10^{-2} cm^{-1}. Hartree-Fock and density functional theory (DFT) methods were the theoretical approaches selected for predicting the geometries and energies of the metal complexes, as they have been implemented in the Gaussian 09W quantum mechanical program [14,15,16]. In this manuscript, we review ^{57}Fe Mössbauer findings and computational results of Fe complexes with several biomaterials.

Chapter 2

EXPERIMENTAL METHODS

Chitosan decamer (Wako, min. 80% deacetylation, molecular weight 1612 Da, and water content max. 10%) and polymer (Sigma chemicals 99.9 % purity level, min. 85% deacetylation, molecular weight range: 310 - 375 kDa) were used without further purification. 0.2 g of chitosan decamer or polymer and anhydrous ferric chloride were mixed in the stoichiometric ratio in 200 ml of deionized water. Preparation of the Fe-complexes with the polymer, water-soluble polymer, and the monomeric forms of chitin and chitosan were undertaken in an acetic acid medium for the polymers and in aqueous medium for the monomer and chondritin sulfate. The chemicals glucosamine, N-acetyl glucosamine, gluconic acid, sulfonated N-acetyl glucosamine, chitin, chitosan, and chondritin sulfate were purchased from WAKO Bioproducts. In one set of experiments, a known amount of a given biopolymer was dissolved in acetic acid. The pH of the solution was adjusted to the desired level and an appropriate amount of the chloride salt of the metal ion was added to the solution. The mixture was stirred for approximately two hours, filtered, and was dried at 50^0C for approximately 12 hours. If no precipitate was formed at room temperature then the water was removed by warming the sample at approximately 50^0C until the solid was formed.

A Mössbauer sample was prepared in the solid state corresponding to a thickness of ~2 mg/cm^2 of ^{57}Fe in a shallow derlin sample cup with an inner cup sealed and frozen with liquid nitrogen. The Mössbauer spectrometer (Ranger model MS 1200, Ranger drive VA900) was used operating in a constant acceleration mode in a transmission geometry. For low temperature measurements, a Cryo Industries 8CN variable-temperature cryostat model was used with liquid He and the temperatures were measured by the calibrated silicon diode, sensitive to different temperature regions. The zero velocity of the Mössbauer spectra referred to the centroid of the room-temperature spectrum of a metallic Fe foil. The spectra were analyzed using the software WMOSS (WEB Research Co., Edina, MN).

Chapter 3

COMPUTATIONAL METHODS

The calculations were performed using the Hartree-Fock method [14] and density functional theory (DFT) [15], as implemented in the Gaussian 09W quantum mechanical program [16]. B3LYP hybrid density functionals [17] were used for the DFT calculations. In the case of open-shell configurations, unrestricted Hartree-Fock and unrestricted B3LYP methods were employed. The following basis sets provided by Gaussian 09W were selected for the study: STO-3G; 6-311G [18]; 6-311G [19] basis sets with p-polarization functions for H, d polarization functions for C and O, and f polarization functions for Fe [20] [i.e., the 6-311G(f,d,p) basis sets]; and the configuration-consistent triple-zeta (cc-pVTz) [21] basis set on Fe. Hartree Fock calculations were performed using STO-3G basis sets, and DFT calculations using 6-311G, 6-311G(f,d,p) and cc-pVTz basis sets. The total number of basis functions and primitive-Gaussian functions for the four types of atoms of the study are shown in Table 1.

Table 1. Size of the basis sets. Labels: #BF = number of basis functions; #PF = number of primitive functions

Atom	STO-3G		6-311G		6-311G(f,d,p)		cc-pVTz	
	#BF	#PF	#BF	#PF	#BF	#PF	#BF	#PF
H	1	3	3	5	6	8		
C	5	15	13	26	18	32		
O	5	15	13	26	18	32		
Fe	18	57	39	71	46	81	68	351

Chapter 4

MÖSSBAUER SPECTROSCOPY

Mössbauer spectroscopy has played a very significant role in the structural elucidation of many Fe-containing metalloenzymes and the technique probes the metal sites directly to obtain information on the oxidation state, the geometry around the metal ion, and its magnetic properties. The Mössbauer parameter isomer-shift (δ_{Fe}), an oxidation state indicator, clearly identifies the number and the nature of ligands that surround the metal ion. For example, an Fe(II) and Fe(III) can be distinguished since they have characteristic values. In addition, Fe(III) in a tetrahedral sulfur environment can be distinguished from an octahedral O/N ligation as the isomer-shifts are distinctly different for these two situations. Another useful and important parameter is the quadrupole splitting (ΔE_Q) which gives a wealth of information regarding the extent of distortion of the coordination sphere. For example, an Fe(III) low-spin state is characterized by a large splitting value, while an Fe(III) high-spin state has a smaller splitting value. In addition, the magnetic properties arising from the presence of unpaired electrons on the transition metal ion offer insight into the nature of metal-metal interactions and illustrate spin-coupling mechanisms that may be present in the Fe-system under study. However, one should bear in mind that these parameters extracted from the Mössbauer spectrum depend on the experimental conditions such as temperature, and the strength of the applied magnetic field [16]. The range of isomer-shift and the quadrupole splitting values obtained, in general, for different Fe-species are listed in Table 2.

Table 2. Typical Mössbauer parameters for Fe compounds
in different oxidation and spin state

	Fe(II)		Fe(III)		Fe(IV)	
	low spin S=0	high-spin S=2	low-spin S=½	high-spin S=3/2	low-spin S=1	high-spin S=2
Isomer Shift (δ_{Fe})	0.4-0.5	1.0-1.4	0.1-0.3	0.4-0.6	0-0.1	0.1-0.3
Quadrupole Splitting (ΔE_Q)	0.7-1.2	2.0-4.0	2.0-3.0	0.5-1.0	1.0	1.0

INTERPRETATION AND ANALYSIS OF THE MÖSSBAUER DATA

It is important to know that the Mössbauer phenomenon rests on the fact that γ radiation can be emitted or absorbed without imparting recoil energy ($E_R = E_\gamma^2/2Mc^2 = 1.95 \times 10^{-3}$ eV, where $E_\gamma = 14.4$ keV and M is the mass of the ^{57}Fe nucleus). In a solid, most of the recoil energy is converted into lattice vibration energy. Mössbauer has shown, however, that there is a certain probability, described by the recoil-free fraction f, that γ emission and absorption take place in solids without recoil. In order to observe the resonance effect, the ^{57}Fe nucleus must be placed in a solid or frozen in a solution matrix. ^{57}Fe is a stable isotope with 2.2% natural abundance. In ^{57}Fe Mössbauer spectroscopy, transitions between the nuclear ground state of ^{57}Fe (nuclear spin $I_g = ½$; nuclear g-factor $g_g = 0.181$) and a nuclear excited state at 14.4 keV ($I_e = 3/2$, $g_e = -0.106$, nuclear quadrupole moment Q) are observed. In the principal axis system of the electric-field-gradient (EFG) tensor, the off-diagonal elements vanish and, since the EFG tensor is traceless, only two independent components need to be specified. By convention, these parameters are V_{zz} and the symmetry parameter η, defined by $\eta = (|V_{xx}| - |V_{yy}|)/|V_{zz}|$. By choosing the coordinate system such that $|V_{zz}| \geq |V_{yy}| \geq |V_{xx}|$, the asymmetry parameter can be restricted to $0 \leq \eta \leq 1$. The conventional Hamiltonian of the quadrupole interaction for the ^{57}Fe nuclear excited state is

$$H_Q = \frac{eQV_{zz}}{12}\left[I_z^2 - I_e(I_e+1) + \eta(I_x^2 - I_y^2)\right] \quad (1)$$

where e is the electron charge, and I_z, I_x, I_y are spin operators. The interaction due to H_Q splits the nuclear excited state into two degenerate doublets. For $\eta = 0$ these doublets are labeled by the magnetic quantum numbers $\pm^3/_2$ and $\pm^1/_2$; for $V_{zz} > 0$ the $^3/_2$ levels have the higher energy. The two doublets are separated in energy by the quadrupole splitting

$$\Delta E_Q = \frac{eQV_{zz}}{2}\sqrt{1+\frac{\eta^2}{3}} \qquad (2)$$

The nuclear ground and excited states of the ^{57}Fe nucleus has magnetic moments which can interact with a magnetic field **H**. The interaction is described by

$$H = -\mu \bullet H = -g_n \beta_n H \bullet I \qquad (3)$$

where g_n is the nuclear g-factor and β_n the nuclear magneton, and I the nuclear spin. In the absence of quadrupolar interactions, the Hamiltonian splits the nuclear states into equally spaced levels of energy $E(m) = -g_n\beta_n H(m)$. The quadrupolar and magnetic interactions are normally combined. A conventional approach followed in spectroscopic investigations of metalloenzymes is to adopt a spin-Hamiltonian formalism for obtaining the fine and hyperfine parameters [42,43]. This approach appears to offer information pertinent to the electronic and magnetic properties of the metal sites involved in the redox process. This information is vital for understanding the redox states and the characteristics of the metal sites. The general Hamiltonian for such a situation is given by

$$H_e = g_e\beta_e \vec{H} \bullet \vec{S} + D[S_z^2 - S(S+1) + \frac{E}{2D}(S_+^2 + S_-^2)]$$
$$- g_n\beta_n \vec{H} \bullet \vec{I} + \vec{I} \bullet \tilde{A} \bullet \vec{S} + H_Q \qquad (4)$$

where H_Q is the term due to quadrupole interaction; g_e the magnetic moment of the electron; β_e the electron Bohr magneton; $\vec{H}$ the applied magnetic field; $\vec{S}$ the spin of the system; S_z the component of the spin along the z-direction of the reference frame; S_+ and S_- the step-up and step-down operators related to the x and y components of the spin by $S_x = \frac{1}{2}(S_+ + S_-)$ and $S_y = \frac{1}{2i}(S_+ - S_-)$; D and E the axial and rhombic zero-field splitting parameters respectively; g_n and β_n the nuclear g-factor and nuclear Bohr magneton respectively; $\vec{I}$ the nuclear spin; and $\widetilde{A}$ the hyperfine coupling tensor [22]. This Hamiltonian is widely used for most spin-resonance techniques, such as Electron Paramagnetic Resonance, Magnetic Circular Dichroism, and Mössbauer Spectroscopy [23].

Chapter 5

RESULTS AND DISCUSSION

The following complexes have been studied by Mössbauer spectroscopy: Fe with glucose, Fe with cellobiose , Fe with glucosamine, Fe with chitosan, and Fe with chitin. The following complexes have been investigated using computational methods: Fe(II) with glucose, Fe(II) with glucosamine, Fe(II) with protonated glucosamine, and Fe(III) with protonated glucosamine.

I. EXPERIMENTAL RESULTS

(A) Fe-Glucose and Fe-Cellobiose

Cellobiose is the dimer of glucose and the Mössbauer spectra of Fe-glucose and Fe-cellobiose, measured at room temperature, show a simple quadrupole doublet with an isomer shift δ_{Fe} and a quadrupole splitting ΔE_Q of 1.30 mm/s and 2.90 mm/s, respectively. Although the Mössbauer parameters are virtually identical for both complexes, the line width of the Fe-cellobiose complex appears to be broader than that of the Fe-glucose complex. However, both systems exhibit Mössbauer parameters typical of high-spin Fe(II). It is not uncommon to have a large electric field gradient at the Fe nucleus, especially for an Fe(II) high-spin state because of its d^6 electron configuration (see table 2 for the range of values). The presence of such a combination of large δ_{Fe} and ΔE_Q values unambiguously identifies the metal ion to be in a high-spin ferrous state [^{5}D state of Fe(II)]. A least-squares fit of the spectral data results on a line-width Γ (full-width at half maximum – FWHM) of nearly 0.40 mm/s, a value somewhat larger than the natural line-width (0.25 mm/s), suggesting a plausible second species with a slight different ligation but with the same metal oxidation and spin state [24].

(B) Fe-Glucosamine

The Mössbauer spectra of the Fe complexes with water soluble chitosan and with D-glucosamine monomer were measured at 4.2 K and are shown in parts A and B of Figure 2, respectively. A visual examination of the spectrum of the Fe complex with water-soluble chitosan clearly indicates that the spectrum is composed of two components, a magnetic part and a quadrupole doublet, whereas the spectrum of the Fe complex with monomer glucosamine shows a pure quadrupole doublet. Comparison of the line positions of the magnetic spectrum of the Fe-water-soluble chitosan complex with earlier reported data for the Fe(III)-water-insoluble chitosan complex reveals that the magnetic component of the spectra are virtually identical and it accounts for about 35 ± 5% of the total intensity of the spectrum. After removal of this component, the remaining central doublet yields a set of Mössbauer parameters δ_{Fe} = 1.39 mm/s and ΔE_Q = 2.89 mm/s. The spectral data of the Fe-glucosamine-monomer complex, on the other hand, shows a pure quadrupole doublet with an apparent isomer shift δ_{Fe} and quadrupole splitting ΔE_Q of 1.37 and 3.01 mm/s, respectively. The parameters obtained from the least-squares fit of the experimental data of the Fe-glucosamine-monomer complex unequivocally indicate the presence of high-spin Fe(II). On the basis of an analysis of the low-temperature spectral data, it can be concluded that a mixture of Fe(II) and Fe(III) ions is formed when mixing the Fe(II) salt with water-soluble chitosan, while the D-glucosamine complex showed only the presence of Fe(II). It should be pointed out that, in both samples, the Mössbauer parameters are identical within the experimental uncertainties [25,26,27].

To further probe the structural details, Fe complexes of D-glucosamine were prepared at various pH values. Figure 3 shows the Mössbauer spectra of these complexes, measured at 293 K, and at various pH values. These spectra clearly show a quadrupole doublet. Furthermore, the Mössbauer parameters correspond to Fe(II), Fe(III), or Fe(II)/Fe(III), depending on the pH of the reaction mixture. At low pH (1.5-4.8) a clear Fe(II) quadrupole doublet is seen, while at high pH (9.2), a mixture of Fe(II) and Fe(III) species is observed. A further increase of the pH to 11.5 leads to a sole Fe(III) species.

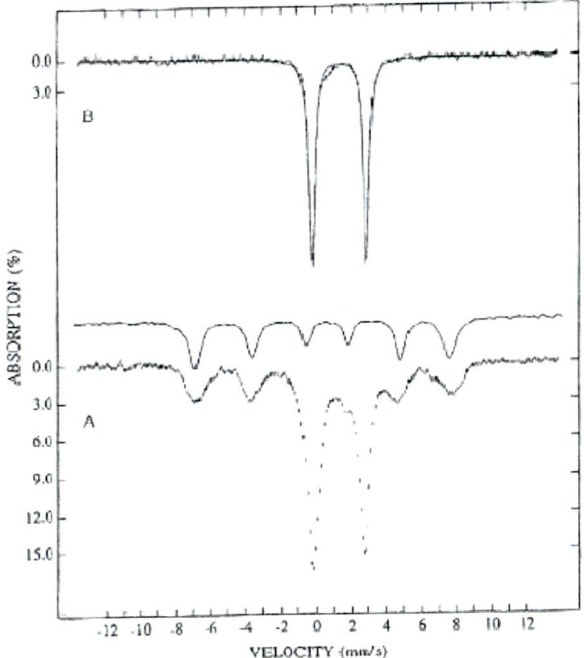

Figure 2. 4.2 K Mössbauer spectra of crystalline powder of (A) Fe-water-soluble chitosan sample. The solid line plotted above the spectrum A is the experimental data of Fe-chitosan polymer complex scaled to 30% of absorption. (B) Fe-glucosamine-monomer sample.

The isomer shift, quadrupole splitting, line width and the corresponding absorption areas determined by a least-squares fit of the spectral data indicate the presence of both Fe(II) and Fe (III) states, depending upon the pH of the reaction mixture. It is interesting to note that a reaction between D-glucosamine and Fe(III) chloride (in an acidic pH range) does not yield an Fe(III) species but shows an Fe(II) species as observed in the Mössbauer spectrum with δ_{Fe} = 1.35 mm/s and ΔE_Q = 3.01 mm/s. This observation is in conformity with the fact that glucosamine acts as a reducing agent. Therefore, it can be concluded that, regardless of the oxidation state of the precursor Fe material, one often ends up with an Fe(II) state if the reaction is maintained in an acidic pH range where glucosamine acts as a reducing agent.

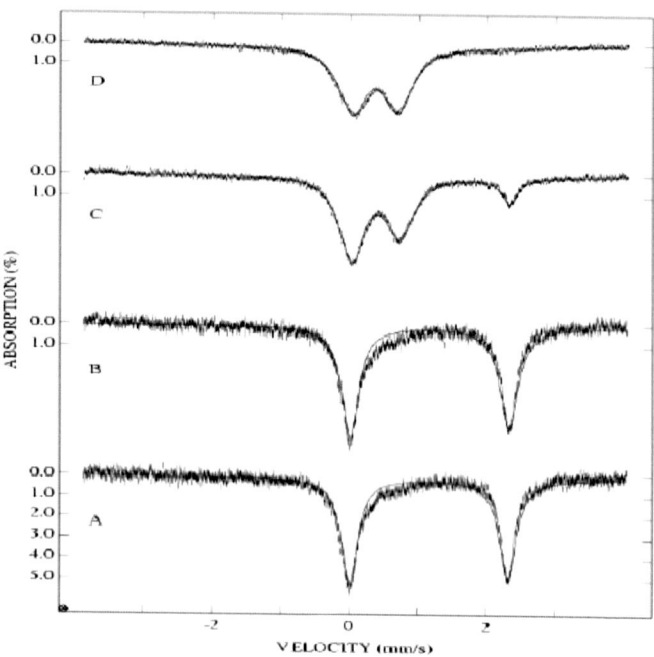

Figure 3. Mössbauer spectra of crystalline powder of Fe-glucosamine samples prepared at different pH values, measured at 293 K. pH values: (A) 1.6; (B) 4.8; (C) 9.2; (D) 11.5.

This important observation should be compared with the fact that a reaction between an iron salt and water-insoluble chitosan always yields an Fe(III)-containing Fe-chitosan complex. This suggests that the coordination properties and the redox potential of the iron sites in complexes with different polymerization degree are the more likely factors that directly influence the oxidation state of the metal ion.

A working structural model for the Fe-chitosan complex has been reported [25] based on the results obtained from the investigations. While the data suggested that there is at least five ligands around Fe in the +3 state, a four coordinated Fe(II) state cannot be completely ruled out. Involvement of the amide nitrogen in the bonding with the metal ion was also proposed. A plausible explanation for the stabilization of the respective oxidation states was offered using simple carbohydrate chemistry, where the presence of β(1-4) glycosidic linkage could be invoked as a possible mechanism.

(C) Fe-Chitosan and Fe-Chitin

The Mössbauer spectra of Fe complexes with chitosan (decamer and polymer), measured at 4.2 K, are shown in Figure 4. As the two spectra are virtually identical and superimposable, the results presented here are applicable to both systems. The spectrum is reminiscent of a classic six line hyperfine signature, with line positions at -6.76, -3.51, -0.51, +1.74, +4.72, and +7.47 mm/s. The Mössbauer spectra measured at 4.2 K, with a magnetic field of 50 mT applied parallel and perpendicular (data not shown) to the incoming γ-rays, are identical, indicating the uniaxial nature of this magnetic system. Any uniaxial system can be characterized by a six-line Mössbauer spectrum with the intensity ratio of 3:2:1:1:2:3 for the six lines respectively [28].

The least-squares fit of the experimental data, the line-width and the shape of each peak clearly indicate the presence of a second species amounting to approximately 10% of the total absorption intensity of the spectrum, which we are currently attributing to an adventitiously bound Fe or the metal ion bonded to the acetylated part of the chitosan (~ max. 20%). However, this spectrum has a magnetic splitting almost identical to the metal ion bonded to the deacetylated part of the chitosan (~ 80%) species and is characteristic of high-spin Fe(III) species. From the line positions obtained by a least-squares fit of the experimental data of the predominant species, and by averaging the line positions, an isomer-shift δ_{Fe} of 0.49 mm/s and an internal magnetic field at the nucleus of 441 kG (H_{int}) were deduced. The quadrupole splitting ΔE_Q can only be indirectly found from this spectrum. The formal relationship between ε and ΔE_Q is

$$\varepsilon = \frac{e^2qQ}{4}\frac{(3\cos^2\theta - 1)}{2} \tag{5}$$

where ε is the energy difference between the lines 1, 2 (Δ_{12}) and 5, 6 (Δ_{56}), and θ is the angle between the principal component of the electric-field-gradient (EFG) tensor and the direction of the internal magnetic field [28]. By comparing the quadrupole splitting obtained from the high-temperature spectral data presented below, it can be inferred that the principal component of the EFG tensor and the direction of the magnetic field are neither perpendicular nor parallel to each other.

These parameters, particularly the isomer-shift δ_{Fe}, are typical of the Fe(III) high-spin state with five or six O/N ligands [29]. An Fe(III) high-spin consists of 5 unpaired electrons (d^5), giving rise to a contribution of 110 kG/unpaired electron. The internal magnetic field in this case is solely due to Fermi-contact interaction, since the orbital angular momentum L is equal to 0. It should be pointed out that the measured internal magnetic field at the Fe nucleus is significantly smaller than any high-spin Fe(III) ion [30]. The difference can only be attributed to the large extent of covalency between the metal and the surrounding ligands. The uniaxial property of the system can be used to parameterize the hyperfine interactions by using a spin-Hamiltonian approach. Using this Hamiltonian, an attempt to simulate the 4.2 K, 50 mT, spectrum yields the parameters $\Delta E_Q = -0.70$ mm/s , $\delta_{Fe} = 0.54$ mm/s, $\eta = -5.0$, hyperfine coupling tensor elements $A_{x,y,z} = -179$ kG, $\Gamma = 0.38$ mm/s, $g_e = 2.0$, $D = -3.0$ cm^{-1}, $E/D = 0.23$, confirming the Fe(III) high-spin state of the Fe-chitosan complex.

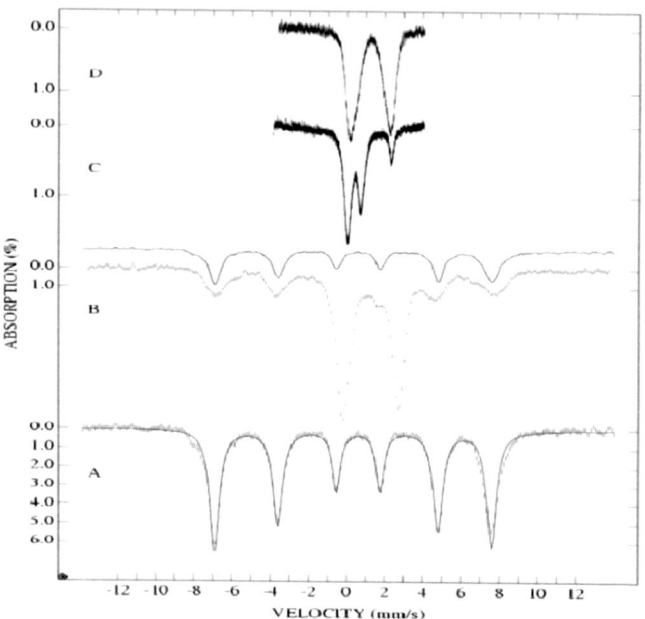

Figure 4. Mössbauer spectra of crystalline powder of (A) Fe-chitosan sample measured at 4.2 K. (B) Fe-water-soluble-chitosan sample measured at 4.2 K. The solid line plotted above spectrum B is the experimental data of the Fe-chitosan-polymer complex, scaled to 30% of absorption. (C) Fe-glucosamine (monomer) at pH 9.2, measured at 293 K, and (D) Fe-chondritin-sulfate complex at pH 4.0 at 293 K.

An analysis of the Mössbauer spectra of the Fe-chitosan complex, measured at different temperatures to extract the isomer shift δ_{Fe} at different temperatures, unambiguously establishes the Fe(III) high-spin state of the complex and supports the validity of the procedure adopted for analyzing the 4.2 K hyperfine spectrum. Our current data show a transition from a magnetic to a paramagnetic phase around 100 K and does not indicate any other structural phase transitions in the temperature range 100 to 300 K. The trend in the isomer shift δ_{Fe} vs temperature is in conformity with expected behavior of a Debye solid. In general, the temperature dependence of the isomer shift δ_{Fe} is due to the second-order Doppler shift and theoretically it is found to be linear if the forces coupling the atoms are assumed to be harmonic. The slope of such a linear plot is given by

$$\frac{d\delta_{Fe}(T)}{dT} = -\frac{3 k E_{\gamma}}{2 M c^2} \quad (6)$$

where c is the velocity of light, M is the atomic mass of the Mössbauer nucleus, k is the Boltzmann constant, and E_γ is the γ-ray energy, which in this case is 14.4 keV [18]. The slope given by the above equation can be calculated to be -7.31×10^{-4} mm/s.K for ^{57}Fe atoms. A plot of isomer shift δ_{Fe} vs temperature for the Fe-chitosan system shows a linear relationship throughout the temperature range studied, with a slope of 6.0×10^{-3} mm/s.K, in reasonable agreement with the predicted value. This suggests that the assumptions under which Equation (6) is derived are valid, i.e., that the forces coupling the atoms in the metal complex are harmonic to a good approximation, and suggests the Debye nature of the solid.

From all the experimental observations and analysis of the data, we proposed a working scheme for the Fe-binding site of chitosan [25], as depicted below in Figure 5.

A similar metal-chitosan complex study has been reported with Cu [31]. It is worthwhile comparing the reported structure with the present results. The proposed structure for the Cu-chitosan complex indicates that Cu is likely to bond to three oxygen atoms and one nitrogen ligand, in a square-planar or a tetrahedral geometry [31]. In this proposed structure, two bonded oxygen atoms and the nitrogen atom are believed to emerge from the monosaccharide group and two such groups are involved in forming the metal coordination sphere. Monterio and Airoldi [31] also argue against another plausible structure in which one bonded oxygen atom and one nitrogen atom arise from the monosaccharide group, while the other two oxygen atoms stem from the water molecules of hydration [31]. From the amino groups contained in chitosan, the amount of water, and the Fe present in the complex, it has been concluded that for each of Fe(III) ion there are 2 moles of amino groups and four moles of oxygen. On the basis of this stoichiometry, the structure $[Fe(H_2O)_{4-x}(Glu)_2Cl_x]Cl_{3-x} \cdot xH_2O$ has been proposed [4], where Glu represents glucosamine.

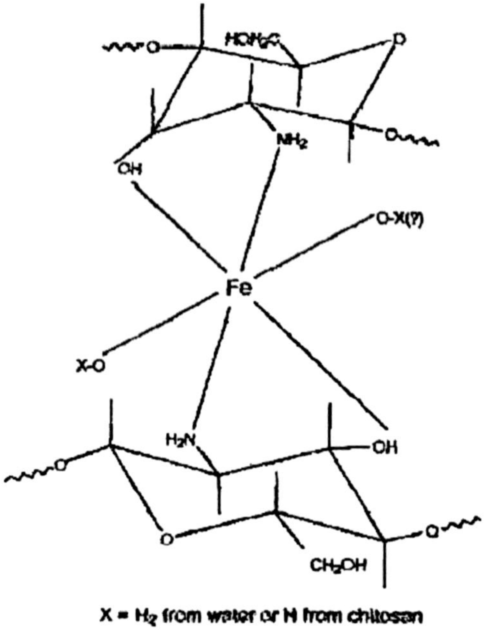

Figure 5. A proposed schematic representation of the Fe-chitosan complex.

Our current study clearly indicates that the Fe-chitosan complex has either a penta- or a hexa-coordinated Fe(III), and it is tempting to speculate that there is at least one molecule of water arising from hydration, and that the remaining N/O ligands are part of two saccharide units of chitosan. This suggests that Fe(III) is indeed coordinated to an amino group (NH_2) of chitosan. The same conclusion can be arrived by comparing the absorptions in the 1500-1700 cm^{-1} region, which are assigned to the bending mode of -NH_2 in chitosan [4a]. The shift of ~200 cm^{-1} for -OH group absorption upon coordination of the Fe(III) to chitosan indicates that Fe(III) is also complexing with -OH groups. Thus, the infrared data indeed lends support and corroborates the proposed working scheme.

II. COMPUTATIONAL RESULTS

Low and high-spin complexes (singlet and quintet multiplicities, respectively) of Fe(II) were studied, with β-D-glucopyranose (glucose) (Complex 1), 2-amino-2-deoxy-β-D-glucose chitosamine (glucosamine) (Complex 2), and protonated glucosamine (Complex 3). A high-spin (hextet multiplicity) was calculated for the Fe(III) complex with protonated glucosamine (Complex 4). Based on Pauling's theory that metal complexes approximate the electronic configuration of noble gases, Fe(II) and Fe(III), with six and five d electrons, respectively, would tend to form hexa-coordinated complexes and thus achieve 6 + 12 electrons (18 electrons) or 5+12 (17 electrons), respectively. The nature of the ligands and the oxidation state of the metal contribute to the crystal-field splitting (Δ_{oct}) of the d orbitals of the transition metal. When Δ_{oct} is large, the complexes have low spin and when it is small, high spin is preferred. Low spin Fe(II) has no unpaired electrons and Fe(III) has one unpaired electron; high spin Fe(II) has four unpaired electrons and Fe(III) has five unpaired electrons [32].

The molecular structures were optimized by minimizing the electronic energy. Computations of the normal vibrational frequencies were carried out to corroborate that stable structures were achieved. Since the model chemistries selected were Hartree-Fock and DFT methods, the force constants for the vibrational analysis were computed analytically. Natural bond orbital analyses (NBO) [33] were performed and the results were compared to Mulliken populations [34]. The nuclear magnetic resonance (NMR) shielding tensors were also computed with the gauge-invariant atomic orbital (GIAO) method [35]. The GaussianView [36] computer program was used as the graphical user interface.

Three types of nuclear interactions are important in Mössbauer spectroscopy: (1) the chemical isomer chemical shift (δ_{Fe}); (2) the electric quadrupole splitting (ΔE_Q); and (3) the hyperfine (or magnetic) splitting [37,38,39]. The theory behind the spectroscopic technique and the relevant parameters have been previously described in the Mösbauer spectroscopy section. In this study, predictions only of the first two parameters were attempted, based on standard curves obtained from experimental values. In previous studies [40], good correlations were obtained between experimental

Mössbauer chemical shifts (δ_{Fe}) and various ways of calculating effective charge on Fe. In particular, Sadoc et al [40 h], correlated δ_{Fe} values with the difference between formal ionic charge and charge calculated using localized atomic orbitals; in that study, charge transfer contribution from the ligands was also considered. In the present study, experimental δ_{Fe} values for five high-spin hexa-coordinated complexes of Fe(II) and Fe(III) were correlated to natural orbital contributions from: (a) the s electrons on Fe, (b) the p and d electrons on Fe, and (c) the sum of the valence p electrons on the ligands multiplied by the formal charge on Fe. Multiplying the ligand p electrons by the formal charge on Fe weighed differently the attraction that the Fe atoms in the compounds have on the ligand p electrons. In a way, performing a multiple linear regression on separate terms allowed for weighing differently the various types of electron density contributions. No relativistic correction was performed on the calculations of electron density; this was the case, both for the molecules used to generate the standard curve, as well as for the molecules for which values of δ_{Fe} were predicted.

The standard curve for predicting electric quadrupole splitting (ΔE_Q) was obtained from experimental values as a function of the EFG, as expressed in Equation 2. The experimental values chosen were all for high-spin, hexa-coordinated, neutral molecules, extrapolated to 0K.

Input Structures

The initial structures contained two β-D-glucopyranose (glucose) monomers (Complex 1), two 2-amino-2-deoxy-β-D-glucose chitosamine (glucosamine) monomers (Complex 2), or two protonated glucosamine monomers (Complex 3 and 4). Glucose and glucosamine are shown in Figure 6. The glucose figure displays the hydroxyl group at C^1 cis to the CH_2OH group attached to C^6 for the β anomer. The O^6-C^6-C^5-O^5 and O^6-C^6-C^5-C^4 torsion angles shown are 60° and 180°, respectively, making them gauche-trans rotamers. The glucosamine figure shows the amine group attached to C^2. For the complex of protonated glucosamine monomers, an additional H atom was attached to N^2; thus, the complex consisted of protonated glucosamine monomer with charge of +1.

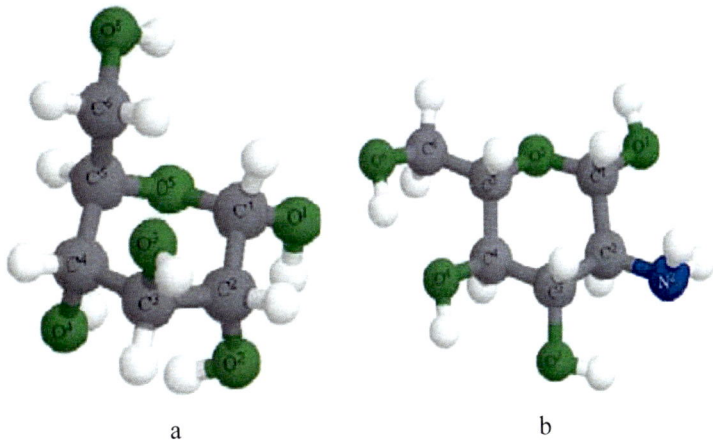

Figure 6. Structures of (a) β-D-glucopyranose (glucose) monomers and (b) β-D-glucose chitosamine (glucosamine) monomers. Green represents the O atoms, gray the C atoms, white the H atoms, and blue the N atom.

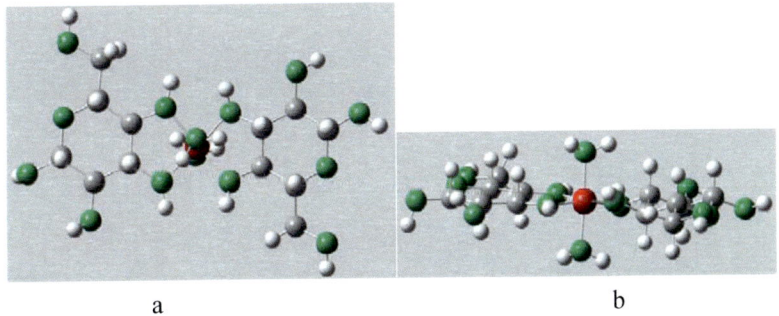

Figure 7. Initial structure of the Fe(II) complex, hexa-coordinated to O^2 and O^3 of two glucose molecules and two H_2O molecules (Complex 1). Green represents the O atoms, gray the C atoms, white the H atoms, and red the Fe atom. (a) top view; (b) side view.

In the complexes, the two monosaccharide molecules were rotated by 180° with respect to each other, both with respect to an axis perpendicular to the main molecular plane and with respect to the molecular plane. In the Fe(II) complexes with glucose and glucosamine, the octahedral Fe atoms were hexacoordinated to the O^3 atoms of the two monosaccharides, to the O atoms of two water molecules placed in the axial position, and to the O^4 atoms of glucose or the N_4 atoms of glucosamine (Complex 1 and 2, respectively). In the Fe (II) and Fe(III) complexes with protonated glucosamine (Complex 3 and 4, respectively), the Fe atoms were coordinated to O^2 and O^3, in addition to the O atoms of two water molecules placed in the axial position. Figure 7 displays the Fe(II) complex with two glucose molecules and two water molecules (Complex 1).

Final Structures

As an example, Figure 8 displays a top view of the final structure of the high spin Fe(II) complex with two glucose molecules and two water molecules (Complex 1), as calculated using B3LYP with 6-311G basis sets. The two water molecules have their H atoms staggered with respect to each other. The two O-Fe distances for the water molecules are both 2.02 Å. The four hydroxyl groups that complex with the Fe have their H atoms away from the Fe atom. The two O^3-Fe distances are 2.16 Å, and the two O^4-Fe distances are 2.14 Å.

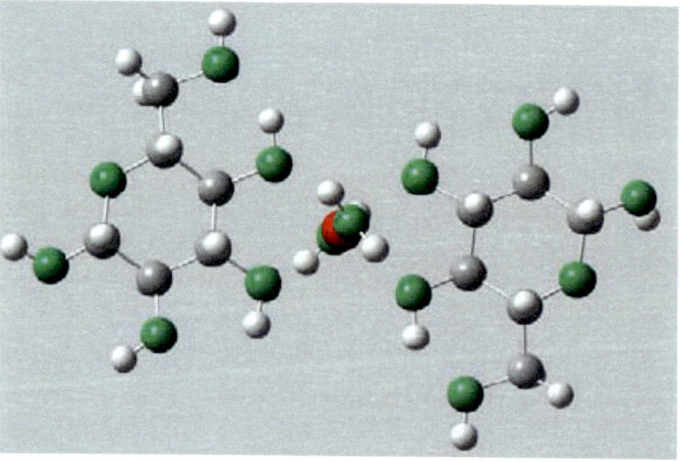

Figure 8. Top view of the high-spin Fe(II) complex with two β-D-glucopyranose (glucose) molecules and two H_2O molecules (Complex 1). Green represents the O atoms, gray the C atoms, white the H atoms, and red the Fe atom.

Binding Energies

Table 3 summarizes the 0K energies for the Fe(II) complex with glucose (Complex 1), Fe(II) complex with glucosamine (Complex 2), Fe(II) complex with protonated glucosamine (Complex 3), and Fe(III) complex with protonated glucosamine (Complex 4). In all cases, the iron atom coordinated with two monosaccharide monomers and two molecules of water. The calculations were done at the Hartree-Fock level with STO-3G basis sets, and at the B3LYP level with 6-311G basis sets or with 6-311G(f,d,p) basis sets. High- and low-spin calculations (quintet and singlet multiplicities, respectively) were performed on Complex 1; only high-spin calculations were performed on the Complexes 2 and 3 (quintet multiplicity), and on Complex 4 (hextet multiplicity). The binding energies were obtained from the 0K electronic energy of the following reaction:

$$\text{Fe(II)} + 2 \text{ sugar molecules} + 2 \text{ water molecules} \rightarrow \text{Complex} \quad (7)$$

As can be seen from Table 3: (a) The complexation reactions are exergonic, with binding energies decreasing with increasing level of calculation and with increasing size of basis set. (b) The energy of the quintet is lower than the energy of the singlet, both for the free Fe(II) atoms as well as for the glucose complex (Complex 1), at all levels of calculation. For the DFT calculations, the energy of the quintet is lower by about 130 kJ mol^{-1} for the complex and by 380 kJ mol^{-1} for the free Fe(II) atom, for both 6-311G and 6-311G(f,d,p) basis sets. (c) The DFT calculations on the glucose complexes (Complex 1), shows binding energies for the singlet of 2.0 and 1.8 MJ mol^{-1} for the 6-311G and 6-311G(f,d,p) basis sets, respectively. The binding energies for the quintet are 1.7 and 1.6 MJ mol^{-1} for the two different basis sets, respectively. The binding energies for the two DFT calculations are within 7% for the singlet and 8% for the quintet. (d) In the case of the glucosamine complex (Complex 2), the binding energies of the quintet at the DFT level are 1.7 and 1.6 MJ mol^{-1} for the 6-311G and 6-311G(f,d,p) basis sets, respectively, i.e., within 7% of each other. (e) At the DFT/6-311G(f,d,p) level, the high-spin (quintet) complex of Fe(II) with protonated glucosamine (Complex 3) has the lowest binding energy of -344 kJ mol^{-1}, whereas the binding energy for the high-spin (hextet) Fe(III) with protonated glucosamine (Complex 4) is 1.5 MJ mol^{-1}.

Table 3. 0K electronic energies for hexa-coordinated iron complexes with two water molecules and two monosaccharides: Fe(II) with glucose (Complex 1), Fe(II) with glucosamine (Complex 2), Fe(II) with protonated glucosamine (Complex 3), and Fe(III) with protonated glucosamine (Complex 4). The 0K electronic energies of the free components are also listed. All energies are in atomic units, except when specified otherwise. BE = binding energy

	HF/STO-3G		B3LYP/6-311G		B3LYP/6-311G(f,d,p)	
	Low spin	High spin	Low spin	High spin	Low spin	High spin
Fe(II)	-1247.6908	-1248.0715	-1262.5983	-1262.7430	-1262.5986	-1262.7437
Fe(III)			-1261.3568	-1261.5901	-1261.3576	-1261.5901
Glucose	-674.4781		-687.1418		-687.3704	
Glucosamine	-654.9658					
Protonated glucosamine	-667.8786					
H$_2$O	-74.9608		-76.4159		-76.4474	
Complex 1	-2747.5653	-2747.9494	-2790.4590	-2790.5111	-2790.9289	-2790.9771
Complex 2		-2708.8816		-2750.7861		-2751.2490
Complex 3						-2751.5268
Complex 4						-2750.8074
BE 1	-0.9965	-0.9899	-0.7453	-0.6527	-0.6946	-0.5978
BE 2		-0.9468		-0.6478		-0.6018
BE 3						-0.1311
BE 4						-0.5652
BE 1 (kJ mol^{-1})	-2,591	-2,599	-1,957	-1,714	-1,824	-1,569
BE 2 (kJ mol^{-1})		-2,486		-1,701		-1,580
BE 3 (kJ mol^{-1})						-344
BE 4 (kJ mol^{-1})						-1,484

	Complex 1 - singlet		Complex 1 - quintet				Complex 2 - quintet			
			α spin		β spin		α spin		β spin	
type	pop	ε (au)	Pop	ε (au)	pop	ε (au)	pop	ε (au)	pop	ε (au)
dxy	0.238	-0.293	0.995	-0.541	0.059	-0.221	0.994	-0.536	0.177	-0.244
dxz	1.989	-0.455	0.998	-0.579	0.020	-0.267	0.997	-0.579	0.016	-0.268
dyz	1.988	-0.452	0.998	-0.576	0.016	-0.262	0.997	-0.575	0.010	-0.261
dx2y2	1.974	-0.455	0.992	-0.517	0.974	-0.425	0.991	-0.517	0.871	-0.399
dz2	0.224	-0.289	0.994	-0.574	0.074	-0.270	0.994	-0.576	0.073	-0.272
Total pop.	6.413		4.976		1.143		4.973		1.148	
Δoct (au)	0.163		0.047				0.051			
Δoct (cm-1)	35,820		10,333				11,092			
Δoct (nm)	279		968				902			

	Complex 3 - quintet				Complex 4 - quintet			
	α spin		β spin		α spin		β spin	
	type	Pop	ε (au)	pop	ε (au)	pop	ε (au)	pop	ε (au)
	dxy	0.993	-0.778	0.056	-0.459	0.996	-0.896	0.047	-0.576
	dxz	0.998	-0.813	0.024	-0.503	0.998	-0.928	0.023	-0.618
	dyz	0.998	-0.806	0.013	-0.491	0.998	-0.921	0.013	-0.605
	dx2y2	0.992	-0.751	0.954	-0.656	0.992	-0.868	0.947	-0.771
	dz2	0.995	-0.806	0.099	-0.508	0.993	-0.920	0.107	-0.624
	Total pop.	4.976		1.146		4.977		1.136	
Δoct (au)		0.044				0.041			
Δoct (cm-1)		9,667				9,009			
Δoct (nm)		1,034				1,110			

Crystal Field Splitting

The electron configurations for free Fe(II) is $1s^2\ 2s^2\ 2p^6\ 3s^2\ 3p^6\ 3d^6$ and for free Fe(III) is $1s^2\ 2s^2\ 2p^6\ 3s^2\ 3p^6\ 3d^5$. Natural bond orbital analyses were performed to calculate orbital populations and to estimate the octahedral crystal field splitting (Δ_{oct}). Table 4 displays the electronic occupation obtained for the d orbitals on the iron atom in the four complexes studied. As can be seen, the total d orbital population was close to 6, for all complexes, even for the Fe(III) complex (Complex 4). Complex 4 had a total charge of +5 due to the two -NH_3^+ groups of the protonated glucosamines. Those -NH_3^+ groups accounted for +1.4 of the charge. In addition, the sugar rings in Complex 4 carried +1.2 higher charge than the Fe(II) complexes, and the C_6 group in Complex 4 carried +0.6 higher charge than the Fe(II) complexes. The hydroxyl groups of the rings and the water molecules had very similar charges in all four complexes. Thus, the additional +1 charge of Fe(III) with respect to Fe(II) of Complex 4 became distributed among the atoms in the rings and the C^6 groups.

The Δ_{oct} values were calculated as the difference between average t_{2g} orbital energies and average e_g orbital energies on Fe. In the case of unrestricted open-shell calculations, only the α orbital energies were used to calculate Δ_{oct}.

Chapter 6

PROPERTIES RELATED TO MÖSSBAUER SPECTROSCOPY

Mulliken atomic charges and natural atomic orbital charges provide useful information about the electron environment of the Fe atom, and are listed in Table 5. Table 5 displays the effective atomic charges for the iron and the ligand atoms of the four complexes studied, based on DFT calculations, using the 6-311G basis sets.

The water oxygens (O^w) are more negative than the sugar oxygens (O^s) or the sugar nitrogens (N^s), creating an anisotropy around the Fe atom. The Fe-O^w distances in the glucose complex (Complex 1) are both 2.10 Å; in the glucosamine complex (Complex 2) are 2.14 and 2.18 Å; in the protonated glucosamine complex with Fe(II) (Complex 3) are 2.11 and 2.12 Å, and in the protonated glucosamine complex with Fe(III) (Complex 4) both are 2.13 Å. The two equivalent Fe-O^s distances in the glucose complex are 2.14 and 2.16 Å (Complex 1); in the glucosamine complex (Complex 2) the Fe-O^s distances are 2.13 Å and the Fe-N^s distances are 2.24 Å; in the protonated glucosamine with Fe(II) (Complex 3) are 2.10 and 2.36 Å, and in the protonated glucosamine with Fe(III) (Complex 4), the Fe-O^s distances are 2.13 and 2.16 Å.

Table 6 contains data for eight high- and low-spin hexa-coordinated Fe(II) and Fe(III) complexes and their chemical isomer shift [17h] that were used to obtain a standard curve for predicting δ_{Fe}. A multiple-regression line was obtained for the experimental isomer shift (δ_{Fe}) versus (a) Fe s electron contribution, (b) Fe p and d electron contribution, (c) ligand p-valence electron contribution multiplied by the change in the formal charge on Fe, calculated from the natural charge. Figure 9 displays the experimental and fitted δ_{Fe} values.

Table 5. Atomic charges for iron and its ligand atoms on the high-spin Fe(II) complex with two glucose molecules and two H_2O molecules (Complex 1), the Fe(II) complex with two glucosamine and two H_2O molecules (Complex 2), the Fe(II) complex with two protonated glucosamines and two H_2O molecules (Complex 3), and the Fe(III) complex with two protonated glucosamines and two H_2O molecules (Complex 4). Nat.=Natural atomic orbital charge; Mul.=Mulliken atomic charge

Complex 1	Atomic charges		Complex 2	Atomic charges	
	Nat.	Mul.		Nat.	Mul.
Sugar 1: O^3	-0.85	-0.80	Sugar 1: O^3	-0.84	-0.80
Sugar 2: O^3	-0.85	-0.79	Sugar 2: O^3	-0.84	-0.80
Sugar 1: O^2	-0.82	-0.78	Sugar 1: N^2	-0.92	-0.89
Sugar 2: O^2	-0.82	-0.78	Sugar 2: N^2	-0.92	-0.89
Fe(II)	1.66	1.61	Fe(II)	1.63	1.59
Water 1: O	-0.97	-0.84	Water 1: O	-0.97	-0.84
Water 2: O	-0.97	-0.84	Water 2: O	-0.97	-0.84
Complex 3	Atomic charges		Complex 4	Atomic charges	
	Nat.	Mul.		Nat.	Mul.
Sugar 1: O^3	-0.79	-0.52	Sugar 1: O^3	-0.79	-0.53
Sugar 2: O^3	-0.79	-0.52	Sugar 2: O^3	-0.79	-0.53
Sugar 1: O^4	-0.86	-0.61	Sugar 1: O^4	-0.83	-0.58
Sugar 2: O^4	-0.86	-0.61	Sugar 2: O^4	-0.83	-0.58
Fe(III)	1.64	1.52	Fe(III)	1.66	1.52
Water 1: O	-0.98	-0.60	Water 1: O	-0.99	-0.61
Water 2: O	-0.98	-0.59	Water 2: O	-0.99	-0.61

Table 6. Experimental isomer shift values (δ_{Fe}), change in the formal charge (formal charge minus natural charge) (CC), and calculated electron contributions for hexa-coordinated Fe complexes. Results of the multiple regression on δ_{Fe} values versus (a) Fe s electron contribution, (b) Fe p and d electron contribution, and (c) ligand (L) p electron contributions multiplied by the change in the formal charge on Fe

Species	δ_{Fe}	Fe	Fe natural atomic orbitals					
	mm s^{-1}	CC	s-core	s-val.	p-core	p-val.	d-val.	L.-p-val.
[Fe(H$_2$O)$_6$]$^{-2}$	1.39	-0.84	6.000	0.219	11.999	0.397	6.190	31.205
[Fe(H$_2$O)$_6$]$^{+3}$	0.50	-1.51	6.000	0.273	11.999	0.481	5.713	31.100
[FeF$_6$]$^{-4}$	1.34	-1.67	6.000	1.015	12.000	0.545	5.673	34.685
[FeF$_6$]$^{-3}$	0.48	-1.67	6.000	0.285	12.000	0.526	5.676	34.657
[Fe(CN)$_6$]$^{4-}$	-0.02	-3.57	5.990	0.449	11.992	1.362	7.686	14.685
[Fe(CN)$_6$]$^{3-}$	-0.13	-4.17	5.990	0.455	11.992	1.412	7.242	15.265
[Fe(NH$_3$)$_6$]$^{-2}$	1.23	-1.11	6.000	0.281	11.999	0.552	6.830	26.790
[Fe(CO)$_6$]$^{2+}$	0.15	-3.85	5.991	0.502	11.993	1.501	7.821	12.091

| | Independent Term | Contribution | | |
		s	p-d	L x CC
Regression coefficients	-1.0208	1.3102	-0.2599	0.0364
Coefficient of determination	0.952			
Uncertainty in δ_{Fe} estimate	0.180 mm s^{-1}			

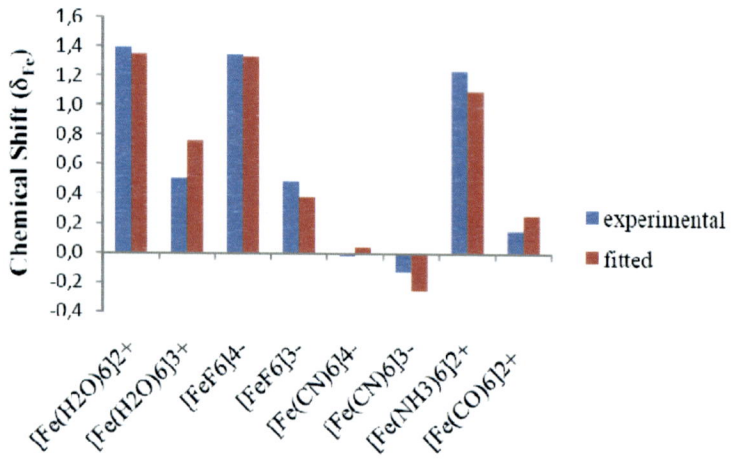

Figure 9. Comparison between experimental **Error! Bookmark not defined.**[h] and calculated chemical isomer shifts (δ_{Fe}) for eight high- and low-spin hexa-coordinated Fe(II) and Fe(III) complexes.

Table 7 summarizes the data used to predicted δ_{Fe} values for the four complexes studied: Complex 1 – Fe(II) with two glucose and two water molecules; Complex 2 – Fe(II) with two glucosamine and two water molecules; Complex 3 – Fe(II) with two protonated glucosamine and two water molecules; Complex 4 – Fe(III) with two protonated glucosamine and two water molecules. The change in the formal charge of iron was taken as the difference between the calculated natural orbital charge and the formal charge. But, the formal charge for all four complexes was taken to be +2. In the case of Complex 4, where the Fe had an expected formal charge of +3, it was still taken as +2, because the third positive charge of the iron became distributed within the sugar rings and the C_6 groups of the protonated glucosamines. Please see the discussion under the section Crystal field splitting.

The following hexa-coordinated, high spin, neutral iron complexes were selected to obtain a standard curve for predicting electric quadrupole splitting ΔE_Q: $FeCl_2(H_2O)_4$, $FeBr_2(H_2O)_4$, and $Fe(HCOO)_2(H_2O)_2$. These species were selected because their counter-ions are not part of the octahedral arrangement. The experimental values [41] for $FeCl_2(H_2O)_4$, and $Fe(HCOO)_2(H_2O)_2$ were extrapolated to 0K using the curves shown in Figure 10. $FeBr_2(H_2O)_4$ was also extrapolated to 0K using the $FeCl_2(H_2O)_4$ parameters.

Properties Related to Mössbauer Spectroscopy

Table 7. Data for the prediction of δ_{Fe} for four high-spin complexes using a standard curve. Complex 1 – Fe(II) with two glucose and two water molecules; Complex 2 – Fe(II) with two glucosamine and two water molecules; Complex 3 – Fe(II) with two protonated glucosamine and two water molecules; Complex 4 – Fe(III) with two protonated glucosamine and two water molecules

	Complex 1	Complex 2	Complex 3	Complex 4
Change in formal charge of Fe (based on natural orbitals)	-0.341	-0.366	-0.357	-0.344*
Multiplicity	quintet	quintet	quintet	hextet
Atomic orbital type	Natural orbital occupancy			
Fe-s-core	5.999	5.999	5.999	5.999
Fe-s-valence	0.212	0.235	0.224	0.221
Fe-p-core	11.998	11.998	11.998	11.998
Fe-p-valence	0.007	0.007	0.006	0.006
Fe-d-valence	6.118	6.121	6.122	6.113
Ligands p-valence	31.049	29.697	31.044	31.016
Predicted δ_{Fe} (mm s^{-1})	2.02 ± 0.18	2.04 ± 0.18	2.02 ± 0.18	2.03 ± 0.18
* Please see discussion in the text.				

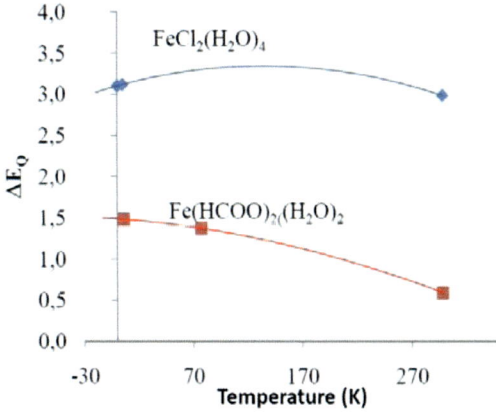

Figure 10. Extrapolation to 0K of experimental values [41] of ΔE_Q for FeCl$_2$(H$_2$O)$_4$, and Fe(HCOO)$_2$(H$_2$O)$_2$.

DFT calculations with a cc-pVTZ basis set on Fe and 6-311g(f,d,p) basis sets on all atoms were performed to obtain values of EFG for FeCl$_2$(H$_2$O)$_4$, FeBr$_2$(H$_2$O)$_4$, and Fe(HCOO)$_2$(H$_2$O)$_2$, as well as for the three molecules studied, in the high spin conformation. Table 9 summarizes the data and Figure 11 displays the experimental values, the fitting curve, and the predicted values of ΔE_Q for the molecules studied.

Table 8. Experimental values [41] of ΔE_Q extrapolated to 0K for $FeCl_2(H_2O)_4$, $FeBr_2(H_2O)_4$, and $Fe(HCOO)_2(H_2O)_2$; predicted values of ΔE_Q (0K) for the following high-spin complexes: Complex 1 – Fe(II) with two glucose and two water molecules; Complex 2 – Fe(II) with two glucosamine and two water molecules; Complex 3 – Fe(II) with two protonated glucosamine and two water molecules; Complex 4 – Fe(III) with two protonated glucosamine and two water molecules

	ΔE_Q (0K) (mm s^{-1})	V_{zz} (au)	η	$V_{zz}(1+\eta^2/3)^{1/2}$ (au)
$FeCl_2(H_2O)_4$	3.10	-2.06	0.07	-2.06
$FeBr_2(H_2O)_4$	2.72	-1.95	0.13	-1.96
$Fe(HCOO)_2(H_2O)_2$	1.49	-2.53	0.14	-2.54
	Predicted ΔE_Q (0K) (mm s^{-1})	V_{zz} (au)	η	$V_{zz}(1+\eta^2/3)^{1/2}$ (au)
Complex 1	2.17 ± 0.46	-2.285	0.118	-2.290
Complex 2	1.01 ± 0.46	-2.743	0.153	-2.754
Complex 3	7.37 ± 0.46	-0.197	0.616	-0.209
Complex 4	3.36 ± 0.46	-1.744	0.487	-1.812

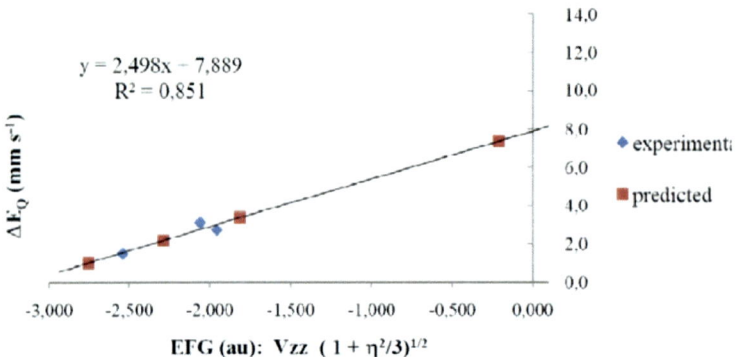

Figure 11. Experimental values [41] of ΔE_Q extrapolated to 0K for $FeCl_2(H_2O)_4$, $FeBr_2(H_2O)_4$, and $Fe(HCOO)_2(H_2O)_2$; predicted values of ΔE_Q (0K) for the following high-spin complexes: Complex 1 – Fe(II) with two glucose and two water molecules; Complex 2 – Fe(II) with two glucosamine and two water molecules; Complex 3 – Fe(II) with two protonated glucosamine and two water molecules; Complex 4 – Fe(III) with two protonated glucosamine and two water molecules.

The large ΔE_Q values predicted is a consequence of considerable deformations of the octahedra. For example, Figure 12 displays the final structure obtained for Complex 1, Fe(II) with glucose and water ligands, and Complex 2, Fe(II) with glucosamine and water ligands. Table 9 lists some angles and dihedral angles of the final structures of the complexes. The atoms have been labeled with two numbers; the one on the left corresponds to the standard atom numbering used in Figure 6, and the one on the right distinguishes between the two monosaccharides. For Complex 1, the distortion of the octahedron occurs through the water molecules, as can be seen from the listed angles in Table 9. For Complex 2, the distortion of the octahedron occurs through the two sugar ligands, whereas the dihedral angles of the water molecule maintain symmetry. The difference between these two complexes is either an NH_2 or a OH group attached to the C^4 of the monosaccharides. The NH_2 group has an additional hydrogen atom that may hinder motion of the water molecules. It is possible that the presence of the extra hydrogen of the NH_2 group forces the two water molecules into very symmetric positions in Complex 2, but in Complex 1 the two water molecules may move with more freedom.

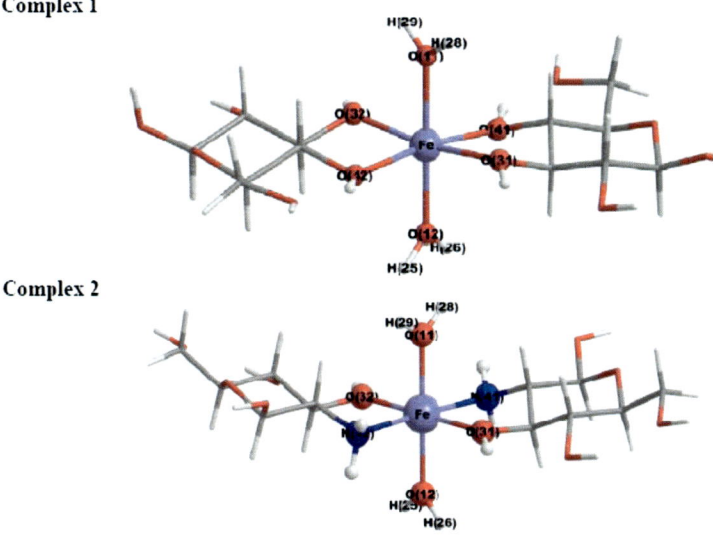

Figure 12. Octahedra distortion in Complex 1 [Fe(II) with glucose and water ligands] and Complex 2 [Fe(II) with glucosamine and water ligands]. Atoms have been labeled with two numbers, the one on the left corresponding to the standard atom numbering used in Figure 6, and the one on the right to distinguish the monosaccharide.

Table 9. Angles and dihedral angles of the final structures of Complex 1 [Fe(II) with two water molecules and two glucose molecules] and Complex 2 [Fe(II) with two water molecules and two glucosamine molecules]. Labels are shown in Figure 12

Complex 1			Complex 2		
	Angle	Dihedral angle		Angle	Dihedral angle
O11-Fe-O42	95		O11-Fe-N42	91	
O11-Fe-O41	95		O11-Fe-N41	91	
O12-Fe-O41	85		O12-Fe-N41	89	
O12-Fe-O42	85		O12-Fe-N42	89	
O11-Fe-O31	85		O11-Fe-O31	90	
O11-Fe-O32	85		O11-Fe-O32	90	
O12-Fe-O31	95		O12-Fe-O31	90	
O12-Fe-O32	95		O12-Fe-O32	89	
O11-Fe-O12	180		O11-Fe-O12	90	
H29-O11-Fe-O12		56	H29-O11-Fe-O12		90
H28-O11-Fe-O12		-123	H28-O11-Fe-O12		90
H25-O12-Fe-O11		-157	H25-O12-Fe-O11		90
H26-O12-Fe-O11		24	H26-O12-Fe-O11		90

Chapter 7

CONCLUSIONS

The following are the principal findings of this investigation.

1. The Mössbauer data of the Fe-incorporated chitosan indicate that the material is magnetic with an unusually small internal field of 441 kG at 4.2 K for a high-spin Fe (III) system, and magnetic spectrum disappears and a pure quadrupole doublet is seen around 100 K.
2. An analysis of the temperature dependence of the isomer-shift suggests that the material behaves like a Debye solid, and the vibration between the atoms is harmonic to a reasonable approximation.
3. Mössbauer data analysis of Fe complexes of chitin, and chitosan, and glucosamine clearly indicates that the metal ion in the 3+ state is stabilized in the case of polymer, while the 2+ state is partially stabilized in the case of water-soluble chitosan and complete stabilization is achieved with D-glucosamine which is the monomer of chitosan. The pertinent information about the oxidation state of the metal ion was obtained by using the isomer-shift deduced from the Mössbauer spectra.
4. Binding energies were computed as the 0K energy difference between the complex and the components (iron ion plus two water molecules, plus two monosaccharides). All binding energies for the four complexes studied by theoretical means were found to be exergonic. The best calculations indicate a binding energy of 1.8 MJ mol^{-1} for low-spin Fe(II)-glucose, 1.6 MJ mol^{-1} for high-spin Fe(II)-glucose, but the high-spin complex was found to be 126 kJ

mol^{-1} more stable than the low-spin complex. The best calculations for high-spin Fe(II)-glucosamine complex resulted in a binding energy of 1.6 MJ mol^{-1}, for Fe(II)-protonated-glucosamine 340 kJ mol^{-1}, and for Fe(III)-protonated-glucosamine 1.5 MJ mol^{-1}.

5. All high-spin complexes were found to have octahedral crystal field splitting (Δ_{oct}) of about 10,000 cm^{-1}. The low-spin Fe(II)-glucose complex had $\Delta_{oct} = 36,000$ cm^{-1}. Iron ion in all complexes had natural atomic charges of about +1.6. The extra charge in Fe(III) became distributed within the sugar rings and the C^6 of the protonated-glucosamines. All high-spin complexes had d^6 configurations.

6. A standard curve based on the δ_{Fe} experimental values of eight Fe complexes was generated, based on a multiple least-squares regression on the following three parameters: (a) s-type natural atomic orbital (NAO) population on Fe; (b) p and d-type NAO population on Fe; (c) p-type NAO populations on the ligands, multiplied by the change in formal charge on the Fe. Using the standard curve for δ_{Fe}, all high-spin complexes studied were predicted to have δ_{Fe} values of 2.0 ± 0.2 mm s^{-1}. The predicted values of δ_{Fe} were within 0.7 mm s^{-1} from the experimental values.

7. Three high-spin hexa-coordinated experimental values of ΔE_Q were used to obtain a standard curve by linear least-squares regression on $V_{zz} (1 + \eta^2/3)^{1/2}$ values. The experimental values were extrapolated to 0K. The predicted values of ΔE_Q for the high-spin complexes were: (2.2 ± 0.5) mm s^{-1} for the Fe(II)-glucose complex; (1.0 ± 0.5) mm s^{-1} for the Fe(II)-glucosamine complex; (7.4 ± 0.5) mm s^{-1} for the Fe(II)-protonated-glucosamine complex; and (3.4 ± 0.5) mm s^{-1} for the Fe(III)-protonated-glucosamine complex. The values suggest that the predicted structure for the Fe(II)-glucosamine complex is much more distorted than the real complex. For the Fe(II)-glucose complex, the distortion occurs through the water molecules, whereas for the Fe(II)-glucosamine complex, much smaller distortions were seen only on the ligand position of sugar portion. The predicted value of ΔE_Q for the Fe(II)-glucose complex was 7% smaller than the experimental value; on the other hand, the predicted ΔE_Q value was 50% smaller for Fe(II)-glucosamine, and was double for Fe(II)-protonated-glucosamine. The discrepancies between the experimental and predicted values will be improved in the future by using standard curves generated from similar chemical systems.

ACKNOWLEDGMENT

Natarajan Ravi acknowledges the Department of Energy (DOE) Grant No. DE-FG52-09NA29518.

REFERENCES

[1] Brine, C. J. In *Chitin, Chitosan and Related Enzymes*; Zikakis, J. P. Ed.; Academic Press: NY 1984.
[2] S.E. Tully, R. Mabon, G.I. Gama, S. Tsai, X. Liu, and Hseieh-Eilson, *J. Am. Chem. Soc.* 126 (2004) 7736.
[3] (a) Ohtakara, A. *Met. Enz.* 1988, 161, 505; Y. Arakai, Y.; Ito, E. *Met. Enz.* 1988, 161, 510; (b) Sanford, P. A. In *Chitin and Chitosan*; G. Skjak-Braek, G.; Anthonsen, T.; Sanford., P., Eds.; Elsevier: Amsterdam 1989.
[4] (a) Nieto, J. M.; Peniche-Covas, C.; Del Bosque, J. *Carb. Pol.* 1992, 18, 221-224; (b) Muzzarelli, R. A. A. In *Chitin*; Pergamon Press; NY 1973; (c) Muzzarelli, R. A. A. In *Natural chelating polymer*; Pergamon Press: Oxford, U.K. 1973; (d) Roberts, G. A. F. In *Chitin Chemistry*, MacMillan, London, 1992.
[5] E.J. Bradbury, L.S.F. Moon, R.J. Popat, V.R. King, G.S. Bennett, P.N. Patel, J.W. Fawcett, S.B. McMohan, *Nature* 416 (2002) 636.
[6] Chiessi, E.; Paradossi, G.; Venanzi, M.; Pispisa, B. *Int. J. Biol. Macro.* 1993, 15, 150.
[7] Gamblin, B. E.; Stevens, J. G.; Wilson, K. L. *Hyp. Int.* 1998, 112, 117.
[8] Mazeau, K.; Winter, W. T.; Chanzy, H. *Macromolecules* 1994, 27, 7606.
[9] Ogawa, K.; Oka, K.; Miyanishi, T.; Hirano, S. In *chitin related enzymes*; Zikakis, J.P. Ed.; Academic Press: Orlando, FL 1984.
[10] (a) Ogawa, K. *Nippon Nogeikagaku Kaishi* 1988, 62, 12225; (b) Schlick, S. *Macromolecules* 1986, 19, 192.
[11] Bernstein, T.; Koetzle, T. F.; Williams, G. J. B.; Meyer, E.; Brice, M. D.; Rodgers, J. R.; Kennard, O.; Shimoaouchi, T.; Tasumi, M. The

Protein Data Bank: A Computer-based Archival File for Macromolecular Structures. *J. Mol. Biol.* 1977, 112, 535.

[12] Chemical Reviews, *Bioinorganic Enzymology*, Holm, R. I.; Solomon, E. Guest Eds.; ACS publication, 1996, 96.

[13] (a) Trautwein, A. X.; Bill, E.; Bominaar, E.; Winkler, H. *Struc. & Bonding* 1991, 71, 1; (b) E. MÜnck, E.; Surerus, K. K.; Hendrich, M. P. *Met. in Enzymology* 1993, 227, 463; (c) Huynh, B. H. *Met. in Enzymology* 1994, 243, 523; (d) J.B. Lynch, C. Juarez-Garcia, E. Münck, L. Que, *J. Biol. Chem.* 264 (1989) 8091; (e) P. Nordlund, H. Eklund, *Curr. Opin. Struc. Biol.* 5 (1955) 758.

[14] Cramer, Christopher J. *Essentials of Computational Chemistry.* Chichester: John Wiley & Sons, Ltd. 2002; pp. 153–189. ISBN 0-471-48552-7.

[15] Parr, R. G.; Yang, W. *Density-functional theory of atoms and molecules.* Oxford Univ. Press, Oxford, 1989. ISBN 0-19-504279-4.

[16] Gaussian 09, Revision A.1, Frisch, M. J., Trucks, G. W., Schlegel, H. B., Scuseria, G. E., Robb, M. A., Cheeseman, J. R., Scalmani, G., Barone, V., Mennucci, B., Petersson, G. A., Nakatsuji, H., Caricato, M., Li, X., Hratchian, H. P., Izmaylov, A. F., Bloino, J., Zheng, G., Sonnenberg, J. L., Hada, M., Ehara, M., Toyota, K., Fukuda, R., Hasegawa, J., Ishida, M., Nakajima, T., Honda, Y., Kitao, O., Nakai, H., Vreven, T., Montgomery, Jr., J. A., Peralta, J. E., Ogliaro, F., Bearpark, M., Heyd, J. J., Brothers, E., Kudin, K. N., Staroverov, V. N., Kobayashi, R., Normand, J., Raghavachari, K., Rendell, A., Burant, J. C. Iyengar, S. S. Tomasi, J. Cossi, M. Rega, Millam, N. J., Klene, M. Knox, J. E., Cross, J. B., Bakken, V., Adamo, C., Jaramillo, J., Gomperts, R. E. Stratmann, O. Yazyev, A. J. Austin, R. Cammi, C. Pomelli, J. W. Ochterski, R. Martin, R. L., Morokuma, K., Zakrzewski, V. G., Voth, G. A., Salvador, P., Dannenberg, J. J., Dapprich, S., Daniels, A. D., Farkas, O., Foresman, J. B., Ortiz, J. V., Cioslowski, J., and Fox, D. J., Gaussian, Inc., Wallingford CT, 2009.

[17] (a) Becke, D. "Density-Functional Exchange-Energy Approximation with Correct Asymptotic Behavior". *Physical Review A38* 1988, 3098-3100. (b) Lee, C.; Yang, W.; Parr, R. G. "Development of the Colle-Salvetti Correlation-Energy Formula into a Functional of the Electron Density". *Physical Review B* 1988, 37, 785-789.

[18] Hehre, W. J.; Ditchfield, R.; Pople, J. A.; "Self-consistent molecular orbital methods. 12. Further extensions of Gaussian-type basis sets for use in molecular-orbital studies of organic molecules"; *J. Chem. Phys.* 56 1972, 2257-2261.

References

[19] Ditchfield, R.; Hehre, W. J.; Pople, J. A.; "Self-consistent molecular orbital methods. 9. Extended Gaussian-type basis for molecular-orbital studies of organic molecules"; *J. Chem. Phys.* 1971, 54, 724-728

[20] Hay, P. J.; "Gaussian basis sets for molecular calculations-representation of 3D orbitals in transition-metal atoms"; *J. Chem. Phys.* 1977, 66, 4377-4384.

[21] Kendall, R. A.; Dunning Jr., T. H.; and Harrison, R. J. "Electron affinities of the first-row atoms revisited. Systematic basis sets and wave functions," *J. Chem. Phys.*, 96 (1992) 6796-806.

[22] Greenwood, N. N.; Gibb, T. C. In *Mössbauer Spectroscopy*, Chapman and Hall Ltd., London 1971.

[23] Abraham, A; Bleany, B. In *Electron Paramagnetic Resonance of Transition ions*, Dover Publications 1970.

[24] Yassin Jeilani, Beatriz H. Cardelino, and Natarajan Ravi (unpublished results)

[25] S. C. Bhatia, and N. Ravi, *Biomacromolecules* 1 (2000) 413.

[26] S.C. Bhatia, and N. Ravi, *Biomacromolecules* 4 (2003) 723.

[27] S.C. Bhatia, B. Cardelino, and N. Ravi, *Hyp. Int.* 165 (2005) 339.

[28] Kundig, W. *Nucl. Instr. Met.* 1967, 48, 219.

[29] Moura, I.; Tavares, P.; Moura, J. J. G.; Ravi, N.; Huynh, B. H.; Liu, M-Y.; Le Gall, J. *J. Biol. Chem.* 1992, 267, 4489.

[30] Tavares, P.; Ravi, N.; Moura, J. J. G.; LeGall, J.; Huang, Y-H.; Crouse, B. R.; Johnson, M. K.; Huynh, B. H.; Moura, I. *J. Biol. Chem.* 1994, 269, 10504.

[31] Monteiro, O. C., Jr.; Airoldi, C. *J. Coll. Int. Sci.* 1999, 212, 212.

[32] Wiberg, E.; Wiberg, N.; Holleman, A. F. *Inorganic chemistry*. Academic Press, San Diego, CA, 2001, p. 1180. ISBN 0-12-352651-5.

[33] (a) NBO Version 3.1; Glendening, E. D.; Reed, A. E.; Carpenter, J. E.; Weinhold, F.; QCPE Bull 1990, 10, 58. (b) Foster, J. P.; Weinhold, F.; "Natural hybrid orbitals"; J.Amer.Chem.Soc. 1980, 102, 7211-7218. (c) Reed, A. E.; Curtiss, L. A.; Weinhold, F.; "Intermolecular interactions from a natural bond orbital, donor-acceptor viewpoint"; *Chem.Rev.* 1988, 88, 899-926.

[34] (a) Mulliken, R. S.; "Electronic Population Analysis on LCAO—MO Molecular Wave Functions. I"; *J. Chem. Phys.* 1955, 23, 1833-1840. (b) Mulliken, R. S.; "Criteria for the Construction of Good Self-Consistent-Field Molecular Orbital Wave Functions, and the Significance of LCAO-MO Population Analysis"; *J. Chem. Phys.* 1962, 36, 3428-3440.

[35] Ditchfield, R. "Self-consistent perturbation theory of diamagnetism. I A gauge-invariant LCAO method for N.M.R. chemical shifts." *Mol. Phys.* 1974, *27*, 789-807.
[36] Gaussian View version 4.1.2. Gaussian Inc.
[37] Gütlich, P. "The Principle of the Mössbauer Effect and Basic Concepts of Mössbauer Spectrometry". http://pecbip2.univ-lemans.fr/~moss/webibame/. Last accessed 03/30/2010.
[38] Debrunner, P. G. "Mössbauer spectroscopy of iron porphyrins". In "Iron Porphyrins". Lever, A. B. P.; Gray, H. B.; editors. VCH Publishers: New York, 1989; Vol. 3, pp.137-234.
[39] Zhang, Y.; Mao, J.; Godbout, N.; Oldfield, E. "Mössbauer Quadrupole Splittings and Electronic Structure in Heme Proteins and Model Systems: A Density Functional Theory Investigation". *J. Am. Chem. Soc.* 2002, *124*, 13921-13930.
[40] (a) Blomquist, J.; Roos,B.O.; Sundbom, M. "Interpretation of the ^{57}Fe Isomer Shift by Means of Atomic Hartree–Fock Calculations on a Number of Ionic States". *J. Chem. Phys.* 1971, *55*, 141-145. (b) Duff, K. J."Calibration of the isomer shift for ^{57}Fe"; *Phys. Rev. B* 1973, *9*, 66-72. (c) Nieuwpoort, W.C.; Post, D.; van Duijnen, P.Th. "Calibration constant for ^{57}Fe Mössbauer isomer shifts derived from ab initio self-consistent-field calculations on octahedral FeF$_6$ and Fe(CN)$_6$ clusters". *Phys. Rev. B* 1978, *17*, 91-98. (d) Neese, F. "Prediction and interpretation of the ^{57}Fe isomer shift in Mossbauer spectra by density functional theory". *Inorg. Chim. Acta* 2002, *337* 181-192. (e) Zhang, Y.; Mao, J.; Oldfield, E. "^{57}Fe Mössbauer Isomer Shifts of Heme Protein Model Systems: Electronic Structure Calculations". *J. Am. Chem. Soc.* 2002, *124*, 7829-7839. (f) Liu, T.; Lowell, T.; Han, W.-G.; Noodleman, L. "DFT" *Inorg. Chem.* 2003, *42*, 5244-5251. (g) Sinnecker, S.; Slep, L. D.; Bill, E.; Neese, F. "Performance of Nonrelativistic and Quasi-Relativistic Hybrid DFT for the Prediction of Electric and Magnetic Hyperfine Parameters in ^{57}Fe Mössbauer Spectra". *Inorg. Chem.* 2004, *44*, 2245-2254. (h) Sadoc, A.; Broer, R.; de Graaf, C. "CASSCF study of the relation between the Fe charge and the Mössbauer isomer shift"; *Chemical Physics Letters* 2008, *454*, 196–200.
[41] Hoy, G. R.; Barros, F. de S. "Mössbauer studies of inequivalent ferrous ion sites in ferrous formate". *Phys. Rev.* 1965, *139*, A929-A934.

INDEX

#

^{57}Fe Mössbauer spectroscopy, ix, 10

β

β(1→4) linkages, 1
β(1→4)-D glucuronic acid, 1
β(1→3)-N acetyl-D-galactosamine, 1

A

ab initio computational methods, ix
absorption, 10, 15, 17, 19, 21
acetic acid, 1, 5
acetylamide group, 1
acid, 2, 5
active site, 1
amine, 2, 23
amino groups, 3, 20
anisotropy, 31
arthropods, 1
asymmetry, 11
atomic orbitals, 23, 33
atoms, 7, 19, 20, 23, 24, 25, 26, 27, 28, 31, 32, 37, 38, 41, 48

B

bending, 21
binding energies, 26, 27, 42
binding energy, 27, 29, 42
biomaterials, ix, 4
biopolymer, 4, 5
biopolymers, 1
Boltzmann constant, 20
bonding, 4, 17

C

cancer, 1
carbohydrate, 3, 17
carbohydrates, 1
carbon, 1, 2
carboxyl, 2
cellulose, 1, 3
chelating model, 3
chitin, 1, 3, 5, 13, 41, 47
chitinase, 2
chitosan, ix, 1, 2, 3, 5, 13, 14, 15, 16, 17, 18, 19, 20, 21, 41
chitosinase, 2
chondritin sulfate, 1, 5
clusters, 50
collagen, 2
color, iv

compounds, 10, 23
configuration, 7, 13, 22
configurations, 7, 27, 42
conformity, 15, 19
connective tissue, 2
convention, 11
COOH, 2
coordination, 9, 16, 20, 21
copolymer, 2
correlations, 23
covalency, 18
crabs, 1
crystalline, 15, 16, 19
cytochrome, 3
cytochromes, 3

D

damages, iv
data analysis, ix, 41
degenerate, 11
density functional theory, 4, 7, 50
Department of Energy, 45
DFT, i, iii, 4, 7, 22, 27, 31, 37, 50
distortion, 9, 38, 39, 43
distortions, 43

E

editors, 50
electric field, 13
electron, 11, 12, 13, 18, 22, 23, 27, 31, 33
Electron Paramagnetic Resonance, 3, 12, 49
electrons, 9, 18, 22, 23
elucidation, 9
emission, 10
enzyme chelation, 3
enzymes, 2, 3
EPR, 3
exoskeleton, 1
experimental condition, 9

F

Fe metal complexes, 3
Fe-chitosan complex, ix, 16, 18, 19, 21
Fermi-contact interaction, ix, 18
ferredoxin, 3
ferric species, ix
ferric state, ix
ferrous ion, 51
fibers, 1
force constants, 22
freedom, 38
frequencies, 22

G

Gaussian 09W quantum mechanical program, 4, 7
geometry, 4, 6, 9, 20
glucose, 1, 3, 13, 22, 23, 24, 25, 26, 27, 29, 31, 32, 35, 36, 37, 38, 39, 42, 43

H

Hamiltonian, ix, 11, 12, 18
Hartree-Fock, 4, 7, 22, 26
hemerythrin, 3
hemoglobin, 3
hybrid, 7, 49
hydrogen, 1, 38
hydrogen bonds, 1
hydrogenase, 3
hydroxyl, 3, 23, 25, 28
hydroxyl groups, 25, 28
hyperfine interaction, 4, 18

I

ideal, 1
in transition, 49
insects, 1
interface, 22

internal field, ix, 41
ions, ix, 1, 3, 14, 35, 49
iron, ix, 16, 26, 27, 29, 31, 32, 35, 42, 50
isotope, 10

L

ligand, 20, 23, 31, 32, 33, 43

M

magnetic characteristics, 4
magnetic field, ix, 9, 11, 12, 17, 18
magnetic moment, 11, 12
magnetic properties, 3, 9, 11
magnetic resonance, 3
magnitude, ix
matrix, 3, 10
metal complexes, 1, 3, 4, 22
metal ion, ix, 1, 2, 3, 5, 9, 13, 16, 17, 41
metal ions, 1, 2, 3
metalloenzymes, 1, 3, 9, 11
mixing, 14
molecular structure, 22
molecular weight, 5
molecules, 20, 23, 24, 25, 26, 27, 28, 29, 32, 35, 36, 37, 38, 39, 42, 43, 48, 49
momentum, 18
monomers, 23, 24, 26
monosaccharide, 20, 25, 26, 39
Moon, 47
Mössbauer data, ix, 41
Mössbauer effect, 3
multiple regression, 33

N

N/O ligands, ix, 21
N-deacetylated chitin, 1
neurons, 2
NH_2, 2, 3, 21, 38
nitrogen, 5, 17, 20
nitrogenase, 3

NMR, 3, 22
noble gases, 22
NQR, 3
nuclear magnetic resonance, 22
Nuclear Magnetic Resonance, 3
Nuclear Quadrupole Resonance, 3
nucleus, ix, 10, 11, 13, 17, 18, 20

O

O/N/S ligands, 2
OH, 3, 21, 38
oxidation, 9, 10, 14, 16, 17, 22, 41
oxygen, 20

P

parallel, 1, 17, 18
pendant model, 3
permission, iv
pH, 5, 14, 15, 16, 19
phase transitions, 19
polarization, 7
polymer, 1, 2, 3, 5, 15, 17, 19, 41, 47
polymer chain, 3
polymer chains, 3
polymerization, 16
polymers, 2, 5
polysaccharide, 1
polysaccharide chains, 1
porphyrins, 50
probability, 10
probe, 14
proteins, 2
purification, 1, 5
purity, 5

Q

quantum mechanics, 4

R

radiation, 10
reactions, 27
recommendations, iv
reference frame, 12
regression, 23, 31, 42
regression line, 31
ribonucleotide reductase, 3
rights, iv
rings, 27, 35, 42
room temperature, 5, 13

S

sewage, 1
shape, 17
silicon, 6
simulations, 4
SO_4^{2-}, 2
software, 6
solid state, 5
solubility, 1
species, ix, 10, 14, 15, 17, 35
spectroscopic techniques, 3
spectroscopy, ix, 9, 10, 13, 22, 50
spiders, 1
spin, ix, 9, 10, 11, 12, 13, 14, 17, 18, 19, 22, 23, 25, 26, 27, 29, 30, 31, 32, 35, 36, 37, 38, 41, 42
spinal cord, 2
stabilization, 17, 41
stable complexes, 1
stoichiometry, 20
substrates, 2
sulfate, 1, 2, 5, 19
sulfur, 9
symmetry, 11, 38
synthetic analogues, 3

T

temperature, 6, 9, 14, 18, 19, 20, 41
temperature dependence, 19, 41
theoretical approaches, 4
torsion, 23
Toyota, 48
transition metal, 1, 2, 3, 9, 22
transition metal ions, 2
transmission, 6

V

valence, 23, 31, 36
velocity, 6, 20
vibration, 10, 41

W

wealth, 9